RAND NATIONAL DEFENSE RESEARCH INSTITUTE

T0170135

Cruise Missile Penaid Nonproliferation

Hindering the Spread of Countermeasures
Against Cruise Missile Defenses

Richard H. Speier, George Nacouzi, K. Scott McMahon

Prepared for the Naval Postgraduate School, Project on
Advanced Systems and Concepts for Combating WMD

For more information on this publication, visit www.rand.org/t/RR743

Library of Congress Cataloging-in-Publication Data

Speier, Richard, author.
  Cruise missile penaid nonproliferation : hindering the spread of countermeasures against cruise missile defenses /
Richard H. Speier, George Nacouzi, K. Scott McMahon.
    pages cm
  Includes bibliographical references.
  ISBN 978-0-8330-8770-6 (pbk. : alk. paper)
  1. Arms control. 2. Export controls. 3. Technology transfer. 4. Cruise missile. 5. Antimissile missiles.
I. Nacouzi, George, author. II. McMahon, K. Scott, author. III. Title

JZ5645.S625 2014
327.1'74—dc23   2014034742

Published by the RAND Corporation, Santa Monica, Calif.

© Copyright 2014 RAND Corporation

RAND® is a registered trademark.

Support RAND
Make a tax-deductible charitable contribution at
www.rand.org/giving/contribute

www.rand.org

Preface

The proliferation of weapons of mass destruction (WMD) becomes a greater threat when accompanied by the proliferation of effective means of delivery. Proliferator nations are acquiring the means of delivery—most threateningly, ballistic and cruise missiles. (Ballistic missiles are powered only for the first part of their flight; thereafter, they coast on a ballistic trajectory. Cruise missiles are powered for their entire flight.)

Defenses against missiles can help protect friends and allies from missile attack. The prospect of such protection can reduce the incentive for potential proliferators to acquire WMD and their delivery systems. Once proliferation has occurred, missile defenses can reduce the expected effects of proliferators' forces and thus help deter aggression.

These benefits will be lost, or at least reduced, if proliferators can acquire effective countermeasures against missile defenses. Such countermeasures, when incorporated in an attacker's missile or employed in conjunction with such a missile, are known as penetration aids, referred to here as *penaids*. The subject of this documented briefing is an approach to hindering the proliferation of cruise missile penaids, specifically by adding certain classes of penaid-relevant items and subitems to the Missile Technology Control Regime. A recent study by the same authors developed a similar approach to controlling the proliferation of ballistic missile penaids; the results of that study were published in *Penaid Nonproliferation: Hindering the Spread of Countermeasures Against Ballistic Missile Defenses*, 2014, available at http://www.rand. org/pubs/research_reports/RR378.html.

This documented briefing was prepared in 2013–2014 under the Naval Postgraduate School research task "Cruise Missile Penaid Nonproliferation: New Measures to Dissuade WMD Proliferation and Reinforce Deterrence." It should be of interest to individuals and organizations concerned with missile defense and with missile and WMD nonproliferation.

This research was sponsored by the Defense Threat Reduction Agency and conducted within the International Security and Defense Policy Center of the RAND National Defense Research Institute, a federally funded research and development center sponsored by the Office of the Secretary of Defense, the Joint Staff, the Unified Combatant Commands, the Navy, the Marine Corps, the defense agencies, and the defense Intelligence Community.

For more information on the International Security and Defense Policy Center, see http:// www.rand.org/nsrd/ndri/centers/isdp.html or contact the director (contact information is provided on the web page).

Contents

Summary

This research describes an approach to hindering the spread of countermeasures against cruise missile defenses. (Such countermeasures, when incorporated in an attacker's missile or employed in conjunction with such a missile, are called penetration aids, or *penaids*.) This approach involved compiling an unclassified list of penaid-relevant items that might be subject to internationally agreed-upon export controls.

The list is designed to fit into the export-control structure of the current international policy against the proliferation of missiles capable of delivering weapons of mass destruction. This policy, the Missile Technology Control Regime (MTCR), sets rules agreed to by 34 governments for restricting the export of items, listed in a technical annex.

This report recommends controls on 18 penaid-relevant items and subitems. Because cruise missile penaids can have applications either not restricted by the MTCR (e.g., for manned aircraft) or subject only to the regime's less rigorous controls (e.g., for relatively small cruise missiles), the report recommends that the 18 items be subject to case-by-case export reviews under MTCR procedures. To be effective, these less rigorous controls will require energetic implementation, and cooperation by Russia and China will be critical.

Acknowledgments

Sponsors and Interviews

- **Defense Threat Reduction Agency**

- **Naval Postgraduate School, Project on Advanced Systems and Concepts for Combating WMD**

- **Agencies and contractors interviewed:**
  - **DoS**
  - **DTSA**
  - **OSD**
  - **DoC**
  - **CIA**
  - **NASIC**
  - **MIT Lincoln Lab**
  - **SAIC**
  - **JIAMDO**
    - **Air Force**
    - **Navy**
    - **Army**
  - **Dennis Gormley**

RAND   1

The funds for this research were provided by the Defense Threat Reduction Agency and administered by the Naval Postgraduate School's Project on Advanced Systems and Concepts for Combating WMD.

More than three dozen individuals from the organizations listed in slide 1 and from among the RAND Corporation's own technical specialists provided guidance for this research. The organizations were the U.S. Department of State (DoS), the Defense Technology Security Administration (DTSA), the Office of the Secretary of Defense (OSD), the U.S. Department of Commerce (DoC), the Central Intelligence Agency (CIA), the National Air and Space Intelligence Center (NASIC), the Massachusetts Institute of Technology's Lincoln Laboratory, Science Applications International Corporation (SAIC), armed services contractors serving on the Joint Integrated Air and Missile Defense Organization (JIAMDO), and Dennis Gormley, the most published authority on the threat of cruise missile proliferation.

Individuals from almost all participating organizations attended a January 29, 2014, workshop at RAND to review and comment on preliminary findings. On the basis of suggestions made in the workshop, the authors extensively revised the draft of this report. Thanks for assistance with the research and manuscript preparation go to Christopher Lynch, Gail Kouril, Elizabeth Hammes, and Alex Chinh at RAND. Particular credit is due to Rebecca Gibbons, a RAND Stanton Nuclear Security fellow, for her organization of the workshop and to Jurgen Gobien of SAIC and Ryan Henry of RAND for their careful review of multiple drafts. However, this report reflects only the views of its RAND authors and not necessarily those of any other individuals or any organization.

CHAPTER ONE

Cruise Missile Penaid Nonproliferation

Research Objective

- **Assist U.S. agencies to hinder the spread of countermeasures against cruise missile defenses.**

- **Do this by developing an unclassified list of penaid-relevant items that might be subject to internationally agreed-upon export controls.**

RAND 2

This research was designed to assist U.S. agencies charged with generating policies to discourage the proliferation of weapons of mass destruction (WMD) and cruise missile delivery systems. The objective was to develop new measures to restrict the proliferation of countermeasures (also known, when incorporated in or used with an attacker's missiles, as penetration aids, or *penaids*) against cruise missile defenses.

It is necessary to identify the science and technology underpinning the development of penaids before policies can be designed to control the threat. Therefore, the research team focused on answering the following overarching research question: What technologies and equipment, if proliferated, would constitute an emerging cruise missile penaid threat to the United States, its allies, partners, and others?

A recent study by the same authors developed a similar approach to controlling the proliferation of ballistic missile penaids; the results of that study were published in *Penaid Nonproliferation: Hindering the Spread of Countermeasures Against Ballistic Missile Defenses*, 2014, available at http://www.rand.org/pubs/research_reports/RR378.html.

Research Method

- **Literature review**

- **Interviews**

- **Technical assessment**

- **Workshop**

- **Follow-up literature review and interviews**

RAND

3

The RAND National Defense Research Institute drew on its expertise in the several subjects relevant to the project: U.S. cruise missile defense systems; domestic and foreign development of penetration aids and related technology and equipment; relevant U.S. aerospace systems, technologies, and industry; and related proliferation/nonproliferation matters. RAND analysts conducted a literature review and interviews to identify data sources and solicit the perspectives of leading government and nongovernment experts in subjects relevant to the project. The research team conducted structured interviews and an independent technical assessment to develop a preliminary characterization of the technologies and equipment most critical to the emerging cruise missile penaid threat. Thereafter, the team invited a selected group of experts to participate in a one-day workshop to review the initial characterization of penaid technologies and equipment.

Policy Refinements to the Objective

- **Fullest possible list**

- **MTCR format**

- **Technology, not policy**

- **Be specific or be general**

- **Not considered: Wassenaar Arrangement**

RAND   4

Interviewees for this research and the previous research on ballistic missile penaids suggested several refinements to the project. One was to develop a broad list of penaid-related items from which the most important might later be selected.

Another was to put these items in the format of the Missile Technology Control Regime (MTCR), the international instrument for hindering the spread of WMD-capable missiles. Thirty-four governments currently subscribe to the provisions of the MTCR. The MTCR Annex (MTCR, 2013) is a control list of items presented in a format usable by government export-control officials.

A third suggestion was to focus exclusively on penaid-relevant technology, not on policy questions. This was easier said than done, however. For example, as discussed later, the placement of items on the MTCR Annex determines the degree to which the export of such items will be restricted. Consequently, although this report focuses on technology, it also identifies inescapable policy questions.

Many suggestions focused on a dilemma: whether to be specific enough in describing penaid-related items so that export-control officials could know precisely what to control or whether to discuss such items in more general terms to avoid unintended information transfer to proliferators. The project team used an approach that took into account both concerns by describing specific systems and subsystems in broad terms.

Finally, some interviewees suggested that, in addition to using the MTCR format, the project should use that of the Wassenaar Arrangement, another international regime with a control list (see WA, 2014). In the words of the WA website,

The Wassenaar Arrangement has been established in order to contribute to regional and international security and stability, by promoting transparency and greater responsibility in transfers of conventional arms and dual-use goods and technologies, thus preventing destabilising accumulations.

However, the WA is less restrictive than the MTCR. For that reason, the research team believes that items suggested for the MTCR could be adapted—with a loss of stringency of control—for the WA but that items added to the WA could not be as readily adapted for the MTCR.

Unique Aspects of Cruise Missile Penaids

- **vs. ballistic missiles**
    - **atmospheric operation**
    - **variable flight path**
    - **efficiency for chemical/biological weapon delivery**

- **vs. manned aircraft**
    - **more restricted size, weight, and power**
    - **autonomous operation**

- **vs. UAVs**
    - **interchangeable**
    - **treated identically by the MTCR**

RAND   **5**

The proliferation threat posed by cruise missiles has long been recognized (see Gormley, 2008; McMahon and Gormley, 1995; and Speier and Gormley, 2003). Cruise missiles or unmanned aerial vehicles (UAVs) are the by far the most efficient means of delivering chemical or biological agents because the vehicles can release the agents in a patterned manner. And defenses against cruise missiles are still at a relatively primitive stage of development.

Although the subjects of this report and *Penaid Nonproliferation: Hindering the Spread of Countermeasures Against Ballistic Missile Defenses* are both penaids, features of cruise missile penaids differ in important respects from those of ballistic missile penaids.

The operation of cruise missile penaids is strongly affected by both the atmosphere in which they operate and the effects of gravity on their flight paths and dispersal. In contrast, ballistic missile penaids—except those designed to operate endoatmospherically—are deployed in space with no air resistance and with an apparent lack of gravity (because all elements of the penaids coast on a ballistic trajectory). So, the technologies of cruise missile penaids and ballistic missile penaids are generally quite different.

Moreover, the launch points and flight trajectories of cruise missiles and their penaids are highly variable—indeed, unpredictable—with respect to geography and their orientation vis-à-vis cruise missile defenses. Ballistic missile penaids are generally more predictable in their trajectories.

In contrast to ballistic missiles, which deliver their payloads at point targets, cruise missiles can dispense a liquid or powder payload while flying in a line perpendicular to the prevailing wind. This means that cruise missiles can be far more efficient at delivering chemical or biological agents.

A critical aspect of cruise missile penaids is their similarity to penaids on manned aircraft. Because the MTCR explicitly exempts manned aircraft items from its controls, this presents a problem in differentiating cruise missile penaids. There are some differences, however. Cruise missiles are generally smaller than manned aircraft, meaning that the size, weight, and power of cruise missile penaids will be more limited than that of manned aircraft penaids. In addition, cruise missile penaids operate autonomously, in contrast to the optional crew controls for manned aircraft penaids.

Finally, the relationship of cruise missiles to the broader category of UAVs is critical for defining controls. UAVs, even if ostensibly designed to return to their launch sites, can be used for munitions delivery and one-way missions. The MTCR has been clear on this since the time of its public announcement in 1987, placing identical controls on both uses by setting the tightest restrictions on "complete unmanned aerial vehicle systems (including cruise missile systems, target drones and reconnaissance drones) capable of delivering at least a 500 kg 'payload' to a 'range' of at least 300 km" (MTCR, 2013, Item 1.A.2). In short, the term *UAV* is more inclusive than *cruise missile*, but because both can challenge defenses and be used in attack modes, this report treats cruise missile penaids and UAV penaids, sometimes called "self-protection" (see La Franchi, 2004), interchangeably.

General Problems with a Cruise Missile Penaid Control List

- **Relative immaturity of some of the technology**

- **Utility of some items for manned aircraft and missile defense itself**

- **Tension between being general and being specific**

- **Negotiability**

RAND   6

There are several broad problems with developing a list of items to be controlled.

The first concerns the relative immaturity of cruise missile defense technology compared with ballistic missile defense technology—and the relative immaturity of UAV penaids (or "self-protection") compared with manned aircraft. As these cruise missile-related technologies become more fully developed, new penaid-related items may become apparent.

The second problem concerns the overlap between cruise missile penaids and those on manned aircraft. This overlap necessitates some difficult line-drawing to avoid conflicts between the MTCR's restrictions and the prohibition against interfering with exports related to manned aircraft. This line-drawing is difficult but not impossible. Governments can draw on a variety of information sources to help determine the end use of an export. In addition, as will be explained in greater detail later in this report, targets for testing cruise missile defenses can overlap with (or be indistinguishable from) cruise missile penaids, creating a conflict between permitting defenses, on the one hand, and not proliferating countermeasures to those defenses, on the other.

A third problem, previously noted, is defining the controls on penaid-related items in sufficient specificity to inform export-control personnel without disclosing information that could be helpful to proliferators.

Finally, with any international control list, negotiability is a concern. The research team worked with export-control officials to consider how to address such concerns. The main issue, to be addressed later in this report, is the level of MTCR restrictions to apply to cruise missile penaid-related items.

Candidates for a Control List

- **MTCR Category II**
    1. **complete penaids**
    2. **test targets**
    3. **decoys**
    4. **spectral flares**
    5. **electronic countermeasures**
        a. **DIRCMs**
        b. **DRFM jammers**
        c. **hot clutter**
        d. **electromagnetic pulse**
        e. **communications jammers**
    6. **reactive threat avoidance**
    7. **dispensers/deployment mechanisms**
    8. **maneuverability features**
    9. **software**
    10. **standoff delivery**

- **Possible expansions of current MTCR Category II**
    1. **supersonic/low-altitude avionics**
    2. **anti-jam equipment**
    3. **stealth**
    4. **hardening**

RAND

7

Slide 7 serves as a table of contents for this report. We propose controls on 18 items or subitems and discuss each individually. We propose including these items under the MTCR's less-restrictive category, Category II. Three of these items—electromagnetic pulse generators, air defense communication jammers, and standoff delivery systems—are more speculative, given the problems discussed in Chapter Three.

CHAPTER TWO

The Missile Technology Control Regime

How the MTCR Works

- **Designed to prevent the proliferation of**
  - **Rocket systems or UAVs capable of delivering a 500 kg payload to a range of at least 300 km**

    or
  - **Any rocket system or UAV intended to deliver WMD**

- **Two categories**
  - **Category I – items subject to a strong presumption of export denial**
  - **Category II – items subject to a case-by-case review and no-undercut rule**

- **Enforced by international cooperation and/or U.S. sanctions**

RAND   8

The MTCR seeks to hinder the spread of rockets and UAVs—regardless of purpose (e.g., space launch, reconnaissance)—beyond a specified range/payload capability and sometimes only range capability, or regardless of range or payload capability if the systems are intended to deliver WMD.

The MTCR's Category I list consists of a relatively small number of items subject to the tightest export restrictions. The MTCR Guidelines state that such exports, if they occur at all, must be "rare" and subject to strong provisions with respect to supplier responsibility.

The MTCR's Category II list consists of items that can be used to manufacture Category I items, as well as other missile-related items for potential WMD delivery. Category II items are generally dual-use, applicable to purposes other than those related to Category I items or WMD delivery. So, Category II exports are subject to greater flexibility but nevertheless require case-by-case export reviews and international procedures to avoid under-cutting Category II export denials by MTCR partners.

The MTCR has well-developed procedures for sharing export decision information among its members. The United States has legislation providing for sanctions against domestic and foreign entities that contribute to missile proliferation (see Speier, Chow, and Starr, 2001). In addition, there are United Nations Security Council sanctions, particularly against Iran and North Korea, that proscribe transfers of items in the MTCR Annex.

Key MTCR Items Considered in this Study

- **Category II:**
  - **Items 5, 7, and 8 – Reserved for future use**
  - **Item 10 – Flight control**
  - **Item 11 – Avionics**
  - **Item 15 – Test facilities and equipment**
  - **Item 16 – Modeling, simulation, and design integration**
  - **Item 17 – Stealth**
  - **Item 18 – Nuclear effects protection**
  - **Item 19 – Other complete delivery systems**

RAND

9

As shown on slide 9, of the 18 classes of items in the MTCR Annex's Category II, a number of items could be relevant for cruise missile penaid controls. There are three blank items on the MTCR Annex—Items 5, 7, and 8—that could be used for additional penaid-related hardware or technology.

Items Also Covered by the MTCR

- **Test and production equipment**

- **Materials**

- **Software**

- **Technology**

RAND   10

Each hardware item listed in the MTCR Annex is generally accompanied by a list of related items, shown in slide 10. In particular, design and production technology is treated at least as restrictively as the hardware item itself.

Cruise Missile Penaid Subsystems Potentially Covered by the Current MTCR

- **Item 11 – Avionics**

- **Item 17 – Stealth**

RAND   11

It can be argued that the current (October 17, 2013) version of the MTCR Annex can be interpreted as covering some of the penaid technologies discussed in this report. This applies, in particular, to the Annex items covering avionics and stealth.

However, some penaid-relevant items are not covered—or not explicitly covered. Later in this report, we consider modifications to the existing Annex items.

Of note, there is a series of definitions early in the MTCR Annex. The definition of *payload* includes countermeasure equipment. However, this definition does not constitute a control list. The purpose of the definition is to standardize the calculation of the mass delivered by a missile.

Definitions and Clarifications

- **Cruise Missile Penaid – in this report, countermeasures carried on or with an attacker's cruise missiles to defeat missile defenses**

- **Cruise Missiles, UAVs – in this report, used interchangeably, but *UAVs* is the more inclusive term**

- **Electromagnetic Spectrum – the full radio frequency, infrared, optical, and ultraviolet spectra (i.e., from about 10-nanometer to 100-meter wavelengths)**

RAND   12

Slide 12 defines terms used frequently in this report. When these terms are used, they are meant with the definitions shown.

Category I or Category II?

- **Utility for manned aircraft**

- **Usability for small UAVs = usability for large UAVs**

- **Option to include in Item 5 (currently an open item)**

RAND

13

Because penaids can increase the offensive effectiveness of cruise missiles, it would be desirable to place the tightest MTCR restrictions on penaids for Category I cruise missiles (i.e., those capable of delivering a 500-kg payload to a range of 300 km). However, the project team was unable to develop a general formula to distinguish such items from those with other applications.

First, virtually all cruise missile penaids are at least potentially usable for manned aircraft, an application that the MTCR is explicitly prohibited from controlling.

Second, penaids usable for Category II UAVs (with a payload below 500 kg) are potentially usable for Category I UAVs.

Consequently, a penaid specifically usable only on a Category I UAV would have a mass of 500 kg or greater and would not be usable on a manned aircraft. The project team was unable to find evidence that such a penaid is likely to be developed.

For these reasons, the items described in this report are proposed for the less-restrictive but nevertheless significant MTCR Category II controls. These controls require a case-by-case review to consider the end use of the proposed export—exempting, for example, uses on manned aircraft. Chapter Four describes in more detail how Category II controls are applied.

The decision not to recommend applying Category I controls to cruise missile penaids was the most difficult of this project. Many specialists wanted Category I restrictions applied to one item or another because of their sensitivity. But the project team could find no way to apply these restrictions under the current MTCR.

It should be noted that, if the end use is determined to be WMD delivery, the MTCR Guidelines automatically upgrade the item to Category I restrictions. Moreover, for the same reasons that penaids for less capable UAVs can be used on more capable ones, this report does not recommend limiting penaid controls on the basis of a missile's range. Consequently, the 300-km range Category II requirement, which appears in the MTCR Annex (Item 19) does not constrain the controls we propose.

CHAPTER THREE

Items Proposed for Penaid Export Controls

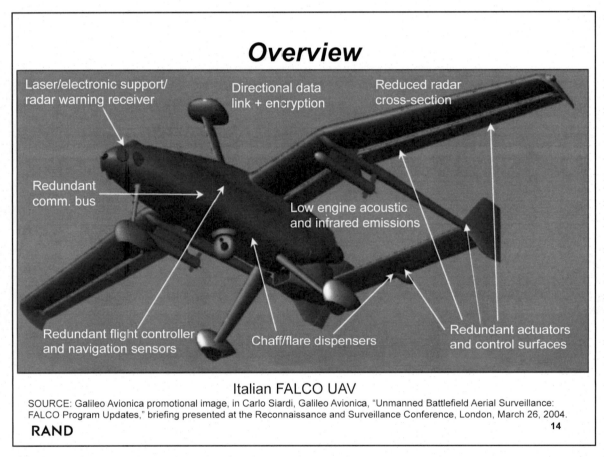

Overview

Laser/electronic support/ radar warning receiver

Directional data link + encryption

Reduced radar cross-section

Redundant comm. bus

Low engine acoustic and infrared emissions

Redundant flight controller and navigation sensors

Chaff/flare dispensers

Redundant actuators and control surfaces

Italian FALCO UAV

SOURCE: Galileo Avionica promotional image, in Carlo Siardi, Galileo Avionica, "Unmanned Battlefield Aerial Surveillance: FALCO Program Updates," briefing presented at the Reconnaissance and Surveillance Conference, London, March 26, 2004.

RAND
14

This report includes illustrations of some of the items suggested for MTCR controls. For example, as shown in slide 14, a given UAV or cruise missile may contain many penaid-related components.

Selectively Restricted Items: Exportable on a Case-by-Case Basis

• **MTCR CATEGORY II**

[Note: These items have alternative uses that should not necessarily be restricted, such as for manned aircraft or missile defenses themselves.]

Countermeasure subsystems and penetration aids designed to saturate, confuse, evade, or suppress missile defenses, and *usable* on *or with* unmanned air vehicles, including:

RAND   15

Slide 15 applies to all candidate MTCR items discussed in this report. Note that some penaid items, such as decoys, might not be carried on the UAV itself but may be deployed by another vehicle or system in conjunction with the operation of the UAV.

1. Complete Penaids

Countermeasures integrated into complete subsystems for installation on UAVs

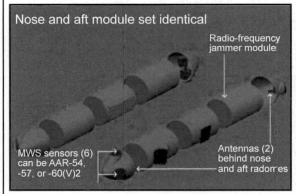

Nose and aft module set identical

Radio-frequency jammer module

MWS sensors (6) can be AAR-54, -57, or -60(V)2

Antennas (2) behind nose and aft radomes

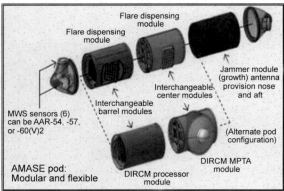

Flare dispensing module

Flare dispensing module

MWS sensors (6) can be AAR-54, -57, or -60(V)2

Interchangeable barrel modules

Interchangeable center modules

Jammer module (growth) antenna provision nose and aft

(Alternate pod configuration)

AMASE pod: Modular and flexible

DIRCM processor module

DIRCM MPTA module

Danish Apache Modular Aircraft Self-Protection Equipment (AMASE)

SOURCE: Terma promotional images.

Issue: The same countermeasures may be used on manned aircraft or on Category I or Category II UAVs. Are Category II case-by-case reviews adequate?

RAND

16

The most obvious candidate for controls is the complete penaid subsystem. Slide 16 shows an integrated penaid on the left and some potential modules on the right.

As discussed earlier, a case-by-case review will be needed to determine whether the complete subsystem is to be used on a UAV or on a manned aircraft. This raises the question of whether MTCR Category II procedures are adequate for dealing with cruise missile penaids. This issue applies to all other penaid items proposed for controls and is discussed in more detail in Chapter Four.

2. Test Targets

Complete test targets or their subsystems simulating countermeasures when observed by missile defense sensors

Serbian rocket-powered test target   Signature control modules

SOURCE: Photos by Miroslav Gyürösi, courtesy of *Jane's Missiles and Rockets*.

Issue: Friends and allies may need test targets for training and exercises.

RAND   17

Missile defense test targets simulate offensive missiles—and, often, penaids. They are used in exercises of missile defense sensor and interceptor systems. Such targets create a number of proliferation problems. For example, their technologies may be indistinguishable from—or, at least, interchangeable with—those of penaids. At a minimum, their development and testing offer a perfect cover for the development and testing of the penaids themselves. Consequently, they should be restricted in the same manner as complete penaids.

There is much legitimate international cooperation in missile defense, however. And missile defense capabilities need to be tested against realistic targets. How can international cooperation occur if participants do not share test targets?

There are options for resolving this dilemma, all of which have precedents in nonproliferation practice. One is for nations receiving missile defenses to develop their own test targets. This has the disadvantage of permitting the development of penaid technology. A second possibility would involve allowing the recipient nation to conduct testing on the supplier's territory to avoid having the supplier export the test targets. This would have the additional advantage of allowing the recipient of missile defense equipment to forgo developing its own targets and testing infrastructure. A third possibility would be for the supplier to maintain jurisdiction or control over the test target while it was in the recipient's territory. Under the MTCR Guidelines, if the supplier retains jurisdiction or control, the movement of an item out of the supplier's territory is not considered a "transfer" (i.e., an export). A fourth, and perhaps most likely, possibility takes advantage of the proposed Category II status of the candidates for controls discussed in this report: Any proposed export of a test target would be carefully reviewed to ensure an acceptable end use and end user.

It should be noted that, if a test target were, in fact, a Category I UAV, it would already be covered by Category I MTCR controls.

2. Test Targets (Cont'd)

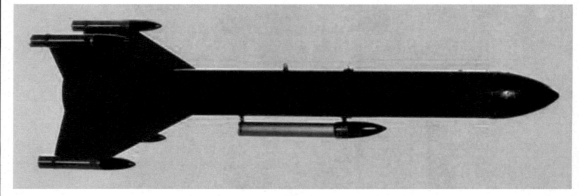

Swedish SM-B6 test target

SOURCE: Enator Miltest AB promotional image.

RAND   18

Slides 17 and 18 illustrate two of the range of test targets on the market—small in slide 17 and large in slide 18. Depending on the sophistication of the test range, virtually all decoys, for example, can be construed to be test targets.

3. Decoys

3. Decoys – towed or free-flying, for launch from or with UAVs

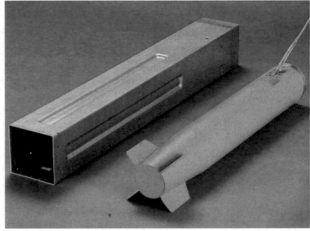

Raytheon towed decoy

SOURCE: Raytheon promotional image.

RAND   19

Decoys can be towed by the cruise missile/UAV or launched by another system. In theory, a decoy could be specifically designed for a Category I UAV, but the research team was unable to define such specificity. However, a decoy not necessarily specifically designed for a Category I system, the Miniature Air Launched Decoy-Jammer, is being integrated for use with a Category I system, the MQ-9 Reaper UAV (see "Miniature Air-Launched Decoy Integrated onto MQ-9 Reaper," 2013; Majumdar, 2014; and Jennings, 2013).

3. Decoys (Cont'd)

3. Decoys – towed or free-flying, for launch from or with UAVs

Israeli free-flying decoy

SOURCE: Israel Military Industries promotional images.

RAND   20

Slide 19 illustrates a towed decoy. Slide 20 illustrates a free-flying decoy, with the full unit shown on the left and alternative configurations on the right.

4. Spectral Flares

Flares designed to match the spectral radiance of the UAV

• UV to IR range

RAND   21

Spectral flares are designed to counter sophisticated heat-seeking defensive missiles that ignore emissions not matching those of the offensive target. Such flares match the emissions spectrum of the cruise missile and mislead the defensive interceptors into targeting the flares rather than the offensive missile.

5. Electronic Countermeasures

Electromagnetic jammers, spoofers, and supporting subsystems designed to counter radar systems, missile seekers, missile fuze sensors, defense electronics, or defense communications – including:

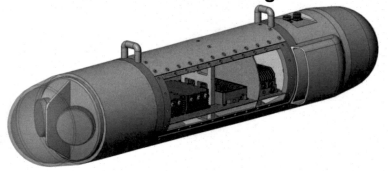

Ukrainian Omut-KM airborne pod-mounted self-defense jammer

SOURCE: Radionix promotional image.

RAND

22

There is a broad range of electronic countermeasures that are usable as penaids. The following slides offer a few examples of relevant current or prospective penaids described in the open literature.

It should be noted that there can be functional overlaps between items for which this report proposes controls. For example, decoys are physical objects designed to mislead the defender, while spoofers are electronic signals with the same ultimate purpose.

Of particular interest is a jamming pod that has been demonstrated for use, not with a cruise missile but with the Category I MQ-9 Reaper UAV (see Carey, 2013).

Electronic Countermeasures (Cont'd)
5.a. DIRCM

Directional infrared countermeasures excepting those specially designed for manned civil aircraft

Northrop Grumman DIRCM

SOURCE: Northrop Grumman promotional image.

RAND   23

Directional infrared countermeasures (DIRCM) are being installed on both military and civil manned aircraft to protect against heat-seeking interceptor missiles (see Military Periscope, 2013a). Their potential use on civil aircraft would be an obvious reason to approve a DIRCM export. Such an exception is consistent with a small number of MTCR exemptions for peaceful uses of missile-related technologies.

Electronic Countermeasures (Cont'd)
5.b. DRFM

Digital radio-frequency memory jammers and spoofers

Mercury Systems digital radio-frequency memory (DRFM)

SOURCE: Mercury Systems promotional image.

Issue: Add analog RFM?

RAND   24

Digital radio frequency memory (DRFM) penaids analyze defensive electronic signals and transmit new signals to counter the defense (see Mercury Systems, undated; the company is one of many manufacturers of these systems). DRFM penaids' export applications must be reviewed with special care.

It is possible to perform DRFM functions with analog, rather than digital, radio frequency memory. We questioned whether it was important to recommend controls on analog radio frequency memory. However, the market has turned decisively in the digital direction.

Electronic Countermeasures (Cont'd)
5.c. Hot Clutter

Subsystems producing hot (actively illuminated) clutter, such as terrain-bounce jammers

Normal radar path from interceptor to UAV and back

Stronger radar from UAV reflects off Earth toward interceptor to deceive it

Terrain-bounce jammer

SOURCE: Adapted from U.S. Patent 5,483,240.

RAND   25

Hot clutter is a false target illuminated in the wavelengths used by the defense to confuse the defending system. It includes chaff dispensed by the cruise missile and actively illuminated to mislead defensive interceptors. Another way to produce hot clutter is with a terrain-bounce jammer, which attempts to mislead the defensive interceptor into detecting a nonexistent target on the earth's surface.

Electronic Countermeasures (Cont'd)
5.d. Electromagnetic Pulse

Explosive and non-explosive generators of non-nuclear electromagnetic pulses or high-power microwaves

 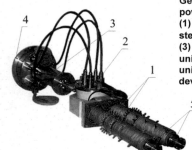

General view of high-power spiral MCG: (1) spiral MCG; (2) pulsed step-up transformer; (3) electric power peaking unit; (4) load-connection unit; (5) MCG triggering devices.

Ukrainian EM pulse generators

SOURCE: Institute for Electromagnetic Research. Used with permission.

RAND   26

Electromagnetic pulse devices generate energy to burn out defender electronics. Boeing and the U.S. Air Force have tested the Counter-Electronics High-powered Microwave Advanced Missile Project (CHAMP; see Jackson, 2012). A Ukrainian firm, the Institute for Electromagnetic Research, advertises such devices. Depending on the design (e.g., an explosive-powered generator) and sophistication, an electromagnetic pulse device may need to be separated from the offensive missile to prevent the offensive missile from committing a self-kill penetration attack. Or the device may be carried in a purpose-built escort UAV. In general, electromagnetic pulse devices are penaids for the future.

Slide 26 shows a non–explosively driven pulse generator on the left and an explosively driven generator on the right.

Electronic Countermeasures (Cont'd)
5.e. Communications Jammers

Onboard or escort flight hardware for jamming cruise missile defense communications, e.g.,

- **Between search radar and tracking radar**

- **Between tracking radar and interceptor missile**

RAND

27

Communications can be a vulnerability of defensive systems, so these systems are usually digital, jam-resistant, and highly directional. However, in the future, penaids might be designed to counter or deny, disable, or disrupt defense-related communications, as opposed to jamming the operations of an interceptor missile.

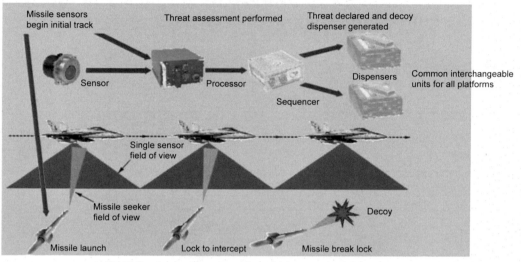

6. Reactive Threat Avoidance

Automated reactive threat avoidance subsystems, exempting collision-avoidance subsystems

BAE Systems sense and avoidance subsystem

SOURCE: BAE Systems promotion image.

RAND

28

Slide 28 illustrates a process that employs several types of hardware. Automated systems can direct an offensive missile to take self-protection actions, such as dispensing decoys or engaging in evasive maneuvers. Reactive threat avoidance systems include both threat-sensing and defense mechanisms. However, such systems have the peaceful application of preventing midair collisions. A Category II review will be needed to distinguish these end uses.

7. Dispensers/Deployment Mechanisms

Dispensers for chaff, obscurants, and flares as well as penetration aid deployment mechanisms for decoys, pulse generators, or precision-guided munitions

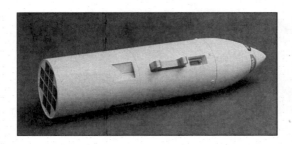 

UK Vicon 78 flare/chaff dispensers

SOURCES: (Left) Thales promotional image; (right) W. Vinten Ltd. promotional image.

RAND   **29**

Penaids of various types may need to be deployed by the offensive vehicle. This requires dispensers or other deployment mechanisms, like those shown in slide 29.

8. Maneuverability Features

Flight hardware, including flight control systems and sensors, designed to operate at acceleration forces greater than 10g sustained

Issue: Precise parameters and their verification

One method of penetrating defenses is to outmaneuver defensive interceptors. Rapid maneuvers involve accelerations (g-forces) that, if too great, cannot be withstood by human pilots. The sustained vertical g-force limit for humans is approximately ten times the force of gravity (i.e., 10g; see the *Wikipedia* entry "G-Force," 2014, for a guide to standard references). Flight hardware for applications above this level can be presumed to be for unmanned aircraft. For example, a Russian offensive missile is reported to be capable of 20–30g maneuvers (see Tikhonov, 2014 [in Russian]).

An issue remains with respect to precisely which g-force parameters to use and how to calculate them.

9. Software

Software and/or algorithms specially designed to enable UAV tactics, techniques, and procedures, such as evasive maneuvers and swarming to penetrate/defeat defensive systems

Issue: Determining functions of software

RAND   31

Sophisticated software may be designed to enable a cruise missile to evade defenses. Such evasive tactics may include maneuvers (see slide 30) or coordination among several cruise missiles to "swarm" a target and overwhelm defenses by force of numbers.

It can be difficult to determine the end use of software. This will be a challenge for case-by-case reviews of Category II software.

10. Standoff Delivery
of Chemical/Biological Weapons

Flight vehicle subsystems specially designed for the standoff delivery of liquid, gases, or powders – including wind velocity profile, turbulence, and inversion layer measurement devices to adjust delivery to environmental conditions

Ukrainian liquid
aerosol dispenser

SOURCE: Institute for Electromagnetic Research.
Used with permission.

Issue: Complex instrumentation, sensors, and algorithms needed for such deployment. Possibly exceeds current state of the art for cruise missile application.

RAND   32

One method for avoiding interception of a chemical or biological agent delivery system is for it to release its payload outside the range of the defender's systems. This candidate item would optimize the delivery of primarily biological agents by adjusting the release of the agents according to local environmental conditions, such as wind direction and the altitude of the inversion layer.

This is a theoretical penaid, however. A few of the many difficulties in developing it include the need for near-real-time mapping of the local environment and onboard modeling and simulation to determine the appropriate deployment tactics. But placing restrictions on this penaid now will help control its proliferation in the future, when the enabling technologies may be more readily available.

Already Restricted on a Case-by-Case Basis, but Might Be Expanded

- **See www.mtcr.info/english/annex.html for fuller descriptions of items on current MTCR control list.**

RAND

At this point, we shift the discussion from items that could be added to the MTCR to items already controlled.

The following discussion concerns four penaid-applicable items already included in the MTCR Annex. This report proposes modifications in these items to more fully address cruise missile penaids.

---

1. Supersonic/Low Altitude Avionics

MTCR Item 11.A.1 (Avionics):

Current text: "Radar and laser radar systems, including altimeters, designed or modified for use in [Category I] systems."

Add precision radar altimeters, forward-looking systems to determine terrain elevation, and guidance, navigation, and control subsystems that enable supersonic or faster flight at altitudes below 30 meters if designed or modified for UAVs.

Issues: Are there any other avionics or supporting sensors that enable autonomous supersonic flight below 30 meters altitude?

RAND   **34**

Avionics and supporting subsystems that permit supersonic or faster flight at very low altitudes can be used preferentially for unmanned systems. A Russian system is reported to be capable of supersonic flight at altitudes of a few meters, and there is a Chinese version of the same system (see "3M-54 Klub," 2014).

The problem will be to distinguish cruise missile–related end uses of such items from uses for manned aircraft.

2. Anti-Jam Equipment

MTCR Item 11.A.3.b.3 (Avionics):

Current text: "Receiving equipment for Global Navigation Satellite Systems (GNSS; e.g., GPS, GLONASS or Galileo) having any of the following characteristics, and specially designed components thereof: . . . b. Designed or modified for airborne applications and having any of the following: . . . 3. Being specially designed to employ anti-jam features (e.g., null steering antenna or electronically steerable antenna) to function in an environment of active or passive countermeasures."

Broaden anti-jam subsystems beyond those for global navigation satellite systems to all sensor, navigation, and communications systems, and add "including multi-mode seekers."

Issue: Add home-on-jam subsystems?

RAND   35

Defensive systems can attempt to jam the electronics of an offensive cruise missile. The offensive countermeasure is anti-jam equipment. The MTCR currently covers only one type of anti-jam equipment. This coverage could be broadened.

One issue is how far to go in controlling anti-jam tactics. Should "home-on-jam" attacks on ground-based defensive systems be covered by the MTCR? Or would this extend MTCR controls too far in the direction of standard offensive tactics to suppress defenses? We raise the issue here for consideration. We recommend, at least initially, keeping MTCR controls focused and not extending them to home-on-jam subsystems.

3. Stealth

MTCR Item 17.A.1 (Stealth):

Current text: "Devices for reduced observables such as radar reflectivity, ultraviolet/infrared signatures and acoustic signatures (i.e., stealth technology). . . ."

Change title of Item 17.A.1 from "Stealth" to "Signature Control." Broaden "reduced" observables to "modified observables." Include communications signature reduction, i.e., "low probability of detection" and "low probability of intercept." Add plasma clouds and active and passive control of electromagnetic signatures (such as emissions from navigation instruments and visual signatures) including signature controls that change but do not reduce the observables.

RAND   36

The current MTCR's stealth item is potentially all-inclusive in terms of devices for reduced observables. It uses the phrase *such as* to make clear that the examples mentioned do not define the limits of the control.

However, the clarity of the item could be improved by adding additional examples, as shown in slide 36. Furthermore, the technology for "reduced" observables can be a subset of the technology for "modified" observables (i.e., signature controls that change, but do not necessarily reduce, the signature of the offensive missile). For that reason, we recommend changing the title of the item to "Signature Control."

4. Hardening

MTCR Item 18.A.1 (Nuclear effects protection):

Current text: "'Radiation Hardened' 'microcircuits' usable in protecting rocket systems and unmanned air vehicles against nuclear effects (e.g., Electromagnetic Pulse (EMP), X-rays, combined blast and thermal effects), and usable for [Category I] systems."

Broaden to "nuclear and non-nuclear effects protection." Include items specially designed for hardening against non-nuclear electromagnetic effects, including non-nuclear electromagnetic pulses, high power microwaves, directed energy thermal radiation, and laser dazzling. Add coverage for items usable for any UAVs.

RAND   **37**

The MTCR Annex already includes items for hardening against nuclear effects. However, it may be appropriate to add other forms of hardening against missile defenses.

CHAPTER FOUR

Implementing Penaid Export Controls

Implementation

- **Review criteria**

- **Assessment of intent**

- **Risk of overloading**

- **Consensus requirements**

RAND

Although penaid export controls present some definitional and structural issues, such issues are familiar matters in the implementation of the MTCR.

For example, the problem of differentiating between appropriate and inappropriate uses of dual-use items is broadly addressed by the MTCR's case-by-case review procedures, which apply the following six criteria (see MTCR, undated):

A. Concerns about the proliferation of weapons of mass destruction;

B. The capabilities and objectives of the missile and space programs of the recipient state;

C. The significance of the transfer in terms of the potential development of delivery systems (other than manned aircraft) for weapons of mass destruction;

D. The assessment of the end use of the transfers, including relevant assurances of the recipient state;

E. The applicability of relevant multilateral agreements;

F. The risk of controlled items falling into the hands of terrorist groups and individuals.

Some of these criteria involve tests of intent (e.g., "objectives," "assessment of the end use," "assurances"). Countermeasure equipment usable for both manned aircraft and cruise missiles (such as chaff dispensers and electronic jammers) can be reviewed for export using the above criteria. Moreover, the process of review involves more than these criteria. Governments can bring to bear information from intelligence, diplomatic, technical, and commercial sources, and they have a variety of means available to influence would-be exporters.

Another concern is the possible negotiating burden if the MTCR is overloaded with up to 18 new or revised items. However, new items can be nested into the control definitions of larger classes of items. One possibility is to fill in the currently empty Item 5 with a new set of subitems, "penetration aids," which could include some of the items suggested in this report.

The greatest difficulty will be obtaining the support of key governments for penaid export controls. The five nations defined by the Nuclear Non-Proliferation Treaty as nuclear-armed states—the United States, the United Kingdom, France, Russia, and China—all have sophisticated penaid programs. Modifications to the MTCR require the consensus of all 34 partners, including Russia but not China (which professes to observe a different form of the MTCR). Russia's agreement is needed for MTCR modification, and China's support is needed to avoid a serious loophole in penaid export controls. The example of Russian and Chinese supersonic low-altitude cruise missiles illustrates how serious this loophole could be. Although both Russia and China are acquiring missile defenses, their current objections to U.S. ballistic missile defenses might spill over into a reluctance to cooperate with controls that would reduce stresses on ballistic or cruise missile defenses.

Concluding Observations

┌───┐

Conclusion

Penaid export controls would help

• Missile nonproliferation

• Missile defense

• Deterrence

But will Russia and China cooperate?

RAND   39

└───┘

This report illustrated how the MTCR Annex could be modified to provide better controls on cruise missile and UAV penetration aids. If enacted, the MTCR modifications suggested here (and those in our previous report on ballistic missile penetration aids; see Speier, McMahon, and Nacouzi, 2014) would constitute one of the most significant adjustments to the regime since its inception in 1987. The recommended modifications, or some variant of them, would strengthen the regime's ability to impede the spread of increasingly lethal cruise missiles capable of penetrating missile defenses and delivering WMD.

Although policy considerations were beyond the scope of this research, moving a complex regime modification to fruition would obviously require careful diplomacy by the United States and like-minded governments. Several government officials interviewed for this study have MTCR-related responsibilities and believe that such an effort would be worthwhile. The recommended MTCR revisions would reinforce the effectiveness of U.S. and allied missile defenses, in turn bolstering protection and deterrence against missile-armed adversaries and enhancing international security.

The largest outstanding question is not the value of restrictions on penaid exports. It is whether Russia and China will support such restrictions.

References

"3M-54 Klub," *Wikipedia*, last updated July 28, 2014. As of April 17, 2014:
http://en.wikipedia.org/wiki/3M-54_Klub

Carey, Bill, "Jamming Pod Demonstrated on MQ-9 Reaper UAV," *AIN Online*, August 23, 2013.

Dexter, Ron, "UAS Combat Threat Survivability," speech delivered at Association for Unmanned Vehicle Systems International Unmanned Systems North America 2011 Conference, Washington, D.C., August 2011.

"FMV SM6," *Jane's Unmanned Aerial Vehicles and Targets*, last updated October 16, 2013.

"G-Force," *Wikipedia*, last updated July 18, 2014. As of April 17, 2014:
http://en.wikipedia.org/wiki/G-force

General Atomics Aeronautical Systems, "GA-ASI Successfully Demonstrates Electronic Attack in USMC Exercise," press release, August 14, 2013. As of July 28, 2014:
http://media.ga.com/2013/08/13/ga-asi-successfully-demonstrates-electronic-attack-in-usmc-exercise

Gormley, Dennis M., *Missile Contagion: Cruise Missile Proliferation and the Threat to International Security*, Westport, Conn.: Praeger Security International, 2008

Gyürösi, Miroslav, "Radionix Ready to Fly Omut-KM Self-Protection Jammer," *Jane's Missiles and Rockets*, November 6, 2012.

———, "Rocket-Powered Target Developed for MANPADS Training," *Jane's Missiles and Rockets*, August 2, 2013.

Haffa, Robert, and Anand Dalta, "Commentary: 6 Ways to Improve UAVs," *Defense News*, March 22, 2012. As of July 28, 2014:
http://www.defensenews.com/article/20120322/C4ISR02/303220009/Commentary-6-Ways-Improve-UAVs

Hoyle, Craig, "Don't fear the Reaper," *The DEW Line*, January 19, 2014. http://www.flightglobal.com/blogs/the-dewline/2014/01/dont-fear-reaper/

Institute for Electromagnetic Research, "Liquid Aerosol Generators," web page, undated(a). As of March 11,2014:
http://iemr.com.ua/liq_aer.htm

———, "Magnetocumulative Generators of High-Power Electric Pulses," web page, undated(b). As of July 28, 2014:
http://www.iemr.com.ua/Figures/mcg_gen_01.jpg

Jackson, Randy, "CHAMP—Lights Out," Boeing Company, October 30, 2012. As of April 23, 2014:
http://www.boeing.com/Features/2012/10/bds_champ_10_22_12.html

Jennings, Gareth, "MALD Decoy to Be Integrated on Predator UAV," *Jane's Missiles and Rockets*, February 14, 2013.

Joint Aircraft Survivability Program Office, *UAV Survivability Enhancement Workshop Report*, proceedings from National Defense Industrial Association Combat Survivability Division workshop, Naval Postgraduate School, Monterey, Calif., November 2004.

La Franchi, Peter, "US Study Recommends Self-Protection for UAVs," *Flight International*, September 2004.

Majumdar, Dave, "Raytheon and General Atomics Team-Up to Integrate MALD onto Reaper," *Flight International*, February 13, 2014.

McLeary, Paul, "Precision Strike: US Army Expanding Soldier Attack Options," *Defense News*, May 6, 2013.

McMahon, K. Scott, and Dennis Gormley, *Controlling the Spread of Land-Attack Cruise Missiles*, Marina Del Rey, Calif.: American Institute for Strategic Cooperation, January 1995.

Mehta, Aaron, "Interview: General Michael Hostage," *Defense News*, February 3, 2014(a).

———, "USAF Leaders Hint at Platforms, Personnel Cuts," *Defense News*, February 3, 2014(b).

———, "Interview: Lt. Gen. Robert Otto," *Defense News*, February 19, 2014(c).

———, "USAF Eyes Future of ISR," *Defense News*, February 19, 2014(d).

Mercury Systems, "Digital RF Memory (DRFM)," web page, undated. As of July 28, 2014:
http://www.mrcy.com/solutiondetail.aspx?id=14946&terms=drfm

Military Periscope, "Apache Modular Aircraft Self-Protection Equipment (AMASE)," web page, last updated December 1, 2012. As of July 28, 2014:
https://www.militaryperiscope.com/weapons/sensors/esm-ew/w0007719.html

———, "ATIRCM (Advanced Threat Infrared Countermeasures)," web page, last updated January 1, 2013a. As of July 28, 2014:
https://www.militaryperiscope.com/weapons/sensors/esm-ew/w0004370.html

———, "Vicon 78," web page, last updated September 1, 2013b. As of July 28, 2014:
https://www.militaryperiscope.com/weapons/sensors/esm-ew/w0002774.html

———, "AN/ALE-50," web page, last updated April 1, 2014. As of July 28, 2014:
https://www.militaryperiscope.com/weapons/sensors/esm-ew/w0003318.html

"Miniature Air-Launched Decoy Integrated onto MQ-9 Reaper," *AirForce Daily*, February 14, 2013.

Missile Technology Control Regime, "Guidelines for Sensitive Missile-Relevant Transfers," web page, undated. As of July 28, 2014:
http://www.mtcr.info/english/index.html

———, "Equipment, Software, and Technology Annex," October 17, 2013. As of July 28, 2014:
http://www.mtcr.info/english/annex.html

MTCR—*See* Missile Technology Control Regime.

Naval Air Systems Command, "NAVAIR Teams Test GPS Anti-Jamming Device on Small UAV," news release, August 8, 2013.

Pappalardo, Joe, "Drones Can Now Jam Enemy Radar," *Popular Mechanics*, November 14, 2013.

Selex ES, "Falco," web page, undated. As of July 28, 2014:
http://www.selex-es.com/-/falco

Siardi, Carlo, Galileo Avionica, "Unmanned Battlefield Aerial Surveillance: FALCO Program Updates," briefing presented at the Reconnaissance and Surveillance Conference, London, March 26, 2004.

Speier, Richard, Brian G. Chow, and S. Rae Starr, *Nonproliferation Sanctions*, Santa Monica, Calif.: RAND Corporation, MR-1285-OSD, 2001. As of July 28, 2014:
http://www.rand.org/pubs/monograph_reports/MR1285.html

Speier, Richard, and Dennis Gormley, "Controlling Unmanned Air Vehicles: New Challenges," Nonproliferation Review, Vol. 10, No. 2, 2003.

Speier, Richard, K. Scott McMahon, and George Nacouzi, *Penaid Nonproliferation: Hindering the Spread of Countermeasures Against Ballistic Missile Defenses*, Santa Monica, Calif.: RAND Corporation, RR-378-DTRA, 2014. As of July 28, 2014:
http://www.rand.org/pubs/research_reports/RR378.html

"Thales Vicon 78 Series 500," *Jane's Unmanned Aerial Vehicles and Targets*, last updated September 2, 2010.

Tikhonov, Sergey, "The Elusive Missile Avenger," *Moscow Ekspert Online* (in Russian), January 27, 2014.

WA—*See* Wassenaar Arrangement.

Wassenaar Arrangement on Export Controls of Conventional Arms and Dual-Use Goods and Technologies, homepage, last updated July 9, 2014. As of July 28, 2014:
http://www.wassenaar.org